# 装载机维修与服务

## （中英双语）

## 工作页

姓名：________________

学号：________________

班级：________________

机 械 工 业 出 版 社

# 工作页目录

# 项目1　装载机介绍与交验

## 任务1.1　装载机介绍

### ☞ 任务描述

通过学习，了解装载机的定义、应用范围、分类、功能、作用。

### ☞ 知识目标

1）了解装载机的定义。
2）了解装载机的分类。
3）掌握装载机的性能参数及各部件名称。

### ☞ 技能目标

1）能列举出生产生活中使用装载机的场景。
2）能分辨出装载机的类别。
3）能编写介绍 CLG856 装载机总体介绍的 PPT。

#### 一、应知应会

装载机的结构组成及各部件名称，装载机基本性能参数。

#### 二、工作过程

**（一）课前准备**

为完成该任务，请检验你是否已掌握以下知识或能力。

1. 请写出装载机的定义。

2. 工作页图 1-1 所示为装载机的总体结构，请写出各部件的名称。

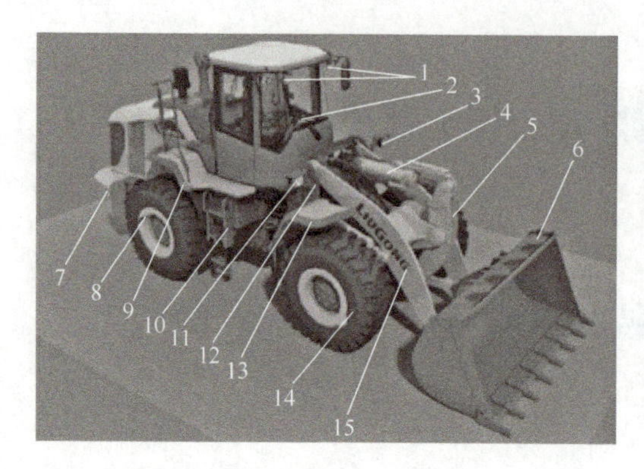

工作页图 1-1 装载机总体结构

1. ________________ 2. ________________

3. ________________ 4. ________________

5. ________________ 6. ________________

7. ________________ 8. ________________

9. ________________ 10. ________________

11. ________________ 12. ________________

13. ________________ 14. ________________

15. ________________ 16. ________________

17. ________________ 18. ________________

19. ________________ 20. ________________

21. ________________

3. 请写出装载机的性能参数。

## （二）计划

1. 小组分工。

工作页表 1-1 小组分工

<table>
<tr><td rowspan="4">小组信息</td><td>班级</td><td colspan="2"></td><td>日期</td><td colspan="2"></td></tr>
<tr><td>小组名称</td><td colspan="2"></td><td>组长</td><td colspan="2"></td></tr>
<tr><td>岗位分工</td><td></td><td></td><td></td><td></td><td></td></tr>
<tr><td>成员</td><td></td><td></td><td></td><td></td><td></td></tr>
</table>

2. 计划讨论。

小组成员共同讨论工作计划，列出本次任务需要做的工作。

### （三）实施

1. 任务实施。

完成 CLG856H 装载机总体介绍的 PPT。

2. 成果分享。

每个小组将实施结果上传到线上教学平台，由 2～3 个小组分别展示和讲解 PPT。

## 三、评价

小组成员各自完成“自我评价”，组长完成“小组评价”，教师完成“教师评价”。

工作页表 1-2 任务评价表

| 序号 | 评价内容 | 自我评价 | 小组评价 | 教师评价 | 分值分配 |
| --- | --- | --- | --- | --- | --- |
| 1 | 遵守安全操作规范 | | | | 5 |
| 2 | 态度端正，工作认真 | | | | 5 |
| 3 | 能进行课前学习，完成任务信息相关练习 | | | | 20 |
| 4 | 能熟练、多渠道查找参考资料 | | | | 5 |
| 5 | 能正确完成任务 | | | | 20 |
| 6 | 方案优化及合理性介绍 | | | | 5 |
| 7 | 能正确回答指导老师的问题 | | | | 15 |
| 8 | 能在规定的时间内完成任务 | | | | 10 |
| 9 | 能与他人团结协作 | | | | 5 |
| 10 | 做好 5S 管理工作 | | | | 10 |
| 合计 | | | | | 100 |
| 拓展项目 | | | | | |
| 总分 | | | | | |

评分说明：

① 评分序号 3 为“课前准备”部分评分分值。

② 总分 = “自我评价” ×20% + “小组评价” ×20% + “教师评价” ×20% + 拓展项目。

③ 如有拓展项目，每完成一个拓展项目，总分加 10 分。

## 四、总结反思

1. 总结学到的新知识。

2. 你对自己在本次任务中的表现是否满意？写出课后反思。

## 五、拓展项目

写出挖掘机各部件的组成及名称，列出清单（请自行附纸）。

# 任务1.2　装载机交验

## ☞ 任务描述

一台新的装载机，需要交付给客户，如何完成交验？

## ☞ 知识目标

1）掌握装载机交机的检查内容及标准。
2）掌握装载机操作与存放注意事项。

## ☞ 技能目标

1）能列出装载机交机检查的内容。
2）能编写CLG856装载机的交验报告。

### 一、应知应会

装载机的交机检查内容及标准。

### 二、工作过程

**（一）课前准备**

为完成该任务，请检验你是否已掌握以下知识或能力。

1. 请列出装载机重点检验的部位。

工作页表1-3　重点检验部位

| 序号 | 验收项目 | 结果 | | 序号 | 验收项目 | 结果 | |
|---|---|---|---|---|---|---|---|
| | | 正常 | 不正常 | | | 正常 | 不正常 |
| 1 | | | | 7 | | | |
| 2 | | | | 8 | | | |
| 3 | | | | 9 | | | |
| 4 | | | | 10 | | | |
| 5 | | | | 11 | | | |
| 6 | | | | 12 | | | |

2. 请说明装载机整机检验的内容，分别怎么检验？

3. 请写出装载机操作与存放的注意事项。

## （二）计划

1. 小组分工。

工作页表 1-4 小组分工

<table>
<tr><td rowspan="4">小组信息</td><td>班级</td><td colspan="2"></td><td>日期</td><td colspan="2"></td></tr>
<tr><td>小组名称</td><td colspan="2"></td><td>组长</td><td colspan="2"></td></tr>
<tr><td>岗位分工</td><td></td><td></td><td></td><td></td><td></td></tr>
<tr><td>成员</td><td></td><td></td><td></td><td></td><td></td></tr>
</table>

2. 计划讨论。

小组成员共同讨论工作计划，列出本次任务需要做的工作。

## （三）实施

1. 任务实施。

完成装载机交验报告。

2. 成果分享。

每个小组将实施结果上传到线上教学平台，由 2 ~ 3 个小组分别展示和讲解 PPT。

# 三、评价

小组成员各自完成“自我评价”，组长完成“小组评价”，教师完成“教师评价”。

工作页表 1-5 任务评价表

| 序号 | 评价内容 | 自我评价 | 小组评价 | 教师评价 | 分值分配 |
|---|---|---|---|---|---|
| 1 | 遵守安全操作规范 | | | | 5 |
| 2 | 态度端正，工作认真 | | | | 5 |
| 3 | 能进行课前学习，完成任务信息相关练习 | | | | 20 |
| 4 | 能熟练、多渠道查找参考资料 | | | | 5 |
| 5 | 能正确完成任务 | | | | 20 |
| 6 | 方案优化及合理性介绍 | | | | 5 |
| 7 | 能正确回答指导老师的问题 | | | | 15 |
| 8 | 能在规定的时间内完成任务 | | | | 10 |
| 9 | 能与他人团结协作 | | | | 5 |
| 10 | 做好 5S 管理工作 | | | | 10 |
| 合计 | | | | | 100 |
| 拓展项目 | | | | | |
| 总分 | | | | | |

评分说明：

① 评分序号 3 为“课前准备”部分评分分值。

② 总分 = “自我评价” ×20% + “小组评价” ×20% + “教师评价” ×20% + 拓展项目。

③ 如有拓展项目，每完成一个拓展项目，总分加 10 分。

## 四、总结反思

1. 总结学到的新知识。

2. 你对自己在本次任务中的表现是否满意？写出课后反思。

## 五、拓展项目

写出挖掘机重点交验检查部位（请自行附纸）。

# 项目 2　装载机维护与保养

## 任务　定期维护与保养计划书的编制

### ☞ 任务描述

通过学习，针对 CLG856H 装载机编制一份 1000h 定期保养计划书。

### ☞ 知识目标

1）了解装载机定期保养的内容与要求。
2）掌握编制装载机定期保养计划的方法。

### ☞ 技能目标

能编写 CLG856H 装载机定期保养内容。

#### 一、应知应会

装载机定期保养内容及要求。

#### 二、工作过程

**（一）课前准备**

为完成该任务，请检验你是否已掌握以下知识或能力。

1. 请写出装载机在工作之前需要检查的项目。

2. 请写出装载机新车走合需要检查的项目。

3. 请写出装载机需要定期维护与保养的项目。

**（二）计划**

1. 小组分工。

工作页表 2-1　小组分工

<table>
<tr><td rowspan="4">小组信息</td><td>班级</td><td colspan="2"></td><td>日期</td><td colspan="2"></td></tr>
<tr><td>小组名称</td><td colspan="2"></td><td>组长</td><td colspan="2"></td></tr>
<tr><td>岗位分工</td><td></td><td></td><td></td><td></td><td></td></tr>
<tr><td>成员</td><td></td><td></td><td></td><td></td><td></td></tr>
</table>

2. 计划讨论。

小组成员共同讨论工作计划，列出本次任务需要做的工作。

**（三）实施**

1. 任务实施。

完成 CLG856H 装载机定期保养计划的编写。

2. 成果分享。

每个小组将实施结果上传到线上教学平台，由 2～3 个小组分别展示和讲解 PPT。

## 三、评价

小组成员各自完成“自我评价”，组长完成“小组评价”，教师完成“教师评价”。

工作页表 2-2　任务评价表

| 序号 | 评价内容 | 自我评价 | 小组评价 | 教师评价 | 分值分配 |
| --- | --- | --- | --- | --- | --- |
| 1 | 遵守安全操作规范 | | | | 5 |
| 2 | 态度端正，工作认真 | | | | 5 |
| 3 | 能进行课前学习，完成项目信息相关练习 | | | | 20 |
| 4 | 能熟练、多渠道查找参考资料 | | | | 5 |
| 5 | 能正确完成任务 | | | | 20 |
| 6 | 方案优化及合理性介绍 | | | | 5 |
| 7 | 能正确回答指导老师的问题 | | | | 15 |
| 8 | 能在规定的时间内完成任务 | | | | 10 |
| 9 | 能与他人团结协作 | | | | 5 |
| 10 | 做好 5S 管理工作 | | | | 10 |
| 合计 | | | | | 100 |
| 拓展项目 | | | | | |
| 总分 | | | | | |

评分说明：

① 评分序号 3 为“课前准备”部分评分分值。

② 总分 = “自我评价” ×20% + “小组评价” ×20% + “教师评价” ×20% + 拓展项目。

③ 如有拓展项目，每完成一个拓展项目，总分加 10 分。

## 四、总结反思

1. 总结学到的新知识。

2. 你对自己在本次任务中的表现是否满意？写出课后反思。

## 五、拓展项目

编制一份挖掘机 1000h 定期保养计划书（请自行附纸）。

# 项目 3　装载机液压系统检修

## 任务 3.1　液压系统工作原理分析

### ☞ 任务描述

通过学习，完成对装载机液压系统的检修。

### ☞ 知识目标

1）掌握先导液压系统控制回路原理。
2）掌握先导液压系统常见故障进行分析并排除。
3）掌握转斗液压系统控制回路原理。
4）掌握转斗液压系统常见故障进行分析并排除。
5）掌握动臂液压系统控制回路原理。
6）掌握动臂液压系统常见故障进行分析并排除。
7）掌握转向液压系统控制回路原理。
8）掌握转向液压系统常见故障并进行分析排除。
9）掌握制动系统控制回路原理。
10）掌握制动系统常见故障并进行分析排除。

### ☞ 技能目标

1）能分析工作液压系统控制回路。
2）能通过控制回路原理，分析常见故障并排除。

#### 一、应知应会

1）装载机液压系统工作原理。
2）装载机液压系统组成。
3）先导液压系统工作原理。
4）转斗液压系统工作原理。
5）动臂液压系统工作原理。
6）转向液压系统工作原理。
7）制动液压系统工作原理。
8）液压系统压力测试。

#### 二、工作过程

**（一）课前准备**

为完成该任务，请检验你是否已掌握以下知识或能力。

1. 请写出装载机液压系统的工作原理。

2. 请写出装载机液压系统的组成。

3. 请写出装载机先导液压系统、转斗液压系统、动臂液压系统的回路。

4. 请写出装载机液压系统压力的测试过程。

5. 请写出装载机转向液压系统压力的测试过程。

6. 请写出装载机制动系统压力的测试过程。

## （二）计划

1. 小组分工。

工作页表 3-1 小组分工

<table>
<tr><td rowspan="4">小组信息</td><td>班级</td><td colspan="2"></td><td>日期</td><td colspan="2"></td></tr>
<tr><td>小组名称</td><td colspan="2"></td><td>组长</td><td colspan="2"></td></tr>
<tr><td>岗位分工</td><td></td><td></td><td></td><td></td><td></td></tr>
<tr><td>成员</td><td></td><td></td><td></td><td></td><td></td></tr>
</table>

2. 计划讨论。

小组成员共同讨论工作计划，列出本次任务需要做的工作。

**（三）实施**

1. 任务实施。

一台装载机出现动臂提升无力问题，经测试，系统压力不足设计值，需要对系统压力测试调压。

2. 成果分享。

每个小组将实施结果上传到线上教学平台，由 2 ~ 3 个小组分别展示和讲解 PPT。

## 三、评价

小组成员各自完成“自我评价”，组长完成“小组评价”，教师完成“教师评价”。

工作页表 3-2 任务评价表

| 序号 | 评价内容 | 自我评价 | 小组评价 | 教师评价 | 分值分配 |
|---|---|---|---|---|---|
| 1 | 遵守安全操作规范 | | | | 5 |
| 2 | 态度端正，工作认真 | | | | 5 |
| 3 | 能进行课前学习，完成任务信息相关练习 | | | | 20 |
| 4 | 能熟练、多渠道查找参考资料 | | | | 5 |
| 5 | 能正确完成任务 | | | | 20 |
| 6 | 方案优化及合理性介绍 | | | | 5 |
| 7 | 能正确回答指导老师的问题 | | | | 15 |
| 8 | 能在规定的时间内完成任务 | | | | 10 |
| 9 | 能与他人团结协作 | | | | 5 |
| 10 | 做好 5S 管理工作 | | | | 10 |
| 合计 | | | | | 100 |
| 拓展项目 | | | | | |
| 总分 | | | | | |

评分说明：

① 评分序号 3 为“课前准备”部分评分分值。

② 总分 = “自我评价” ×20% + “小组评价” ×20% + “教师评价” ×20% + 拓展项目。

③ 如有拓展项目，每完成一个拓展项目，总分加 10 分。

## 四、总结反思

1. 总结学到的新知识。

2. 你对自己在本次任务中的表现是否满意？写出课后反思。

3. 列出你掌握的新技能。

## 五、拓展项目

装载机转向系统压力测试（请自行附纸）。

# 任务 3.2 液压系统性能检测

## 任务描述

通过学习，完成对装载机液压系统性能的测试。

## 知识目标

1）掌握装载机液压缸沉降量的测试方法。
2）掌握装载机液压缸循环时间的测试方法。
3）掌握装载机液压油油温的测试方法。

## 技能目标

1）能按规范要求测量装载机液压缸的沉降量。
2）能按规范要求测量液压缸的循环时间。
3）能够利用测温枪对液压油油温进行测试。

### 一、应知应会

1）装载机液压缸沉降量测试。
2）装载机液压缸循环时间测试。
3）装载机液压油油温测试。

### 二、工作过程

**（一）课前准备**

为完成该任务，请检验你是否已掌握以下知识或能力。

1. 请写出装载机液压缸测试方案。

2. 请写出装载机液压缸循环时间测试步骤。

**（二）计划**

1. 小组分工。

工作页表 3-3 小组分工

<table>
<tr><td rowspan="4">小组信息</td><td>班级</td><td colspan="2"></td><td>日期</td><td colspan="2"></td></tr>
<tr><td>小组名称</td><td colspan="2"></td><td>组长</td><td colspan="2"></td></tr>
<tr><td>岗位分工</td><td></td><td></td><td></td><td></td><td></td></tr>
<tr><td>成员</td><td></td><td></td><td></td><td></td><td></td></tr>
</table>

2. 计划讨论。

小组成员共同讨论工作计划，列出本次任务需要做的工作。

### （三）实施

1. 任务实施。

一台装载机提升到最高后出现掉斗现象，需要对液压缸的沉降量进行测试，以便查找原因。

2. 成果分享。

每个小组将实施结果上传到线上教学平台，由 2 ~ 3 个小组分别展示和讲解 PPT。

## 三、评价

小组成员各自完成“自我评价”，组长完成“小组评价”，教师完成“教师评价”。

工作页表 3-4 任务评价表

| 序号 | 评价内容 | 自我评价 | 小组评价 | 教师评价 | 分值分配 |
| --- | --- | --- | --- | --- | --- |
| 1 | 遵守安全操作规范 | | | | 5 |
| 2 | 态度端正，工作认真 | | | | 5 |
| 3 | 能进行课前学习，完成任务信息相关练习 | | | | 20 |
| 4 | 能熟练、多渠道查找参考资料 | | | | 5 |
| 5 | 能正确完成任务 | | | | 20 |
| 6 | 方案优化及合理性介绍 | | | | 5 |
| 7 | 能正确回答指导老师的问题 | | | | 15 |
| 8 | 能在规定的时间内完成任务 | | | | 10 |
| 9 | 能与他人团结协作 | | | | 5 |
| 10 | 做好 5S 管理工作 | | | | 10 |
| 合计 | | | | | 100 |
| 拓展项目 | | | | | |
| 总分 | | | | | |

评分说明：

① 评分序号 3 为“课前准备”部分评分分值。

② 总分 = “自我评价” ×20% + “小组评价” ×20% + “教师评价” ×20% + 拓展项目。

③ 如有拓展项目，每完成一个拓展项目，总分加 10 分。

## 四、总结反思

1. 总结学到的新知识。

2. 你对自己在本次任务中的表现是否满意？写出课后反思。

3. 列出你掌握的新技能。

## 五、拓展项目

装载机液压油油温测试（请自行附纸）。

# 项目4　装载机电气系统检修

## 任务　主电路工作原理分析与检修

### ☞ 任务描述

通过学习，对装载机主电路工作原理进行分析与检修。

### ☞ 知识目标

1）掌握装载机主电路的结构与功能。

2）掌握装载机主电路主要电器元件。

3）掌握装载机主电路的工作原理与故障分析。

### ☞ 技能目标

1）能正确描述装载机主电路的工作原理。

2）能按规范要求对电源回路进行检修。

#### 一、应知应会

1）装载机主电路的结构与功能。

2）装载机主电路主要电器元件。

3）装载机主电路的工作原理与故障分析。

#### 二、工作过程

**（一）课前准备**

为完成该任务，请检验你是否已掌握以下知识或能力。

1. 请写出装载机电气系统的主要结构。

2. 请写出装载机电气系统的主电路工作原理。

3. 请采用列表法对故障进行分析。

### （二）计划

1. 小组分工。

工作页表 4-1 小组分工

<table>
<tr><td rowspan="4">小组信息</td><td>班级</td><td colspan="2"></td><td>日期</td><td colspan="2"></td></tr>
<tr><td>小组名称</td><td colspan="2"></td><td>组长</td><td colspan="2"></td></tr>
<tr><td>岗位分工</td><td></td><td></td><td></td><td></td><td></td></tr>
<tr><td>成员</td><td></td><td></td><td></td><td></td><td></td></tr>
</table>

2. 计划讨论。

小组成员共同讨论工作计划，列出本次任务需要做的工作。

### （三）实施

1. 任务实施。

一台装载机发电机有阻滞现象和碰刮响声，需要对发电机进行故障排查和分析，并给出解决办法。

2. 成果分享。

每个小组将实施结果上传到线上教学平台，由 2 ~ 3 个小组分别展示和讲解 PPT。

## 三、评价

小组成员各自完成“自我评价”，组长完成“小组评价”，教师完成“教师评价”。

工作页表 4-2 任务评价表

<table>
<tr><td>序号</td><td>评价内容</td><td>自我评价</td><td>小组评价</td><td>教师评价</td><td>分值分配</td></tr>
<tr><td>1</td><td>遵守安全操作规范</td><td></td><td></td><td></td><td>5</td></tr>
<tr><td>2</td><td>态度端正，工作认真</td><td></td><td></td><td></td><td>5</td></tr>
<tr><td>3</td><td>能进行课前学习，完成任务信息相关练习</td><td></td><td></td><td></td><td>20</td></tr>
<tr><td>4</td><td>能熟练、多渠道查找参考资料</td><td></td><td></td><td></td><td>5</td></tr>
<tr><td>5</td><td>能正确完成任务</td><td></td><td></td><td></td><td>20</td></tr>
<tr><td>6</td><td>方案优化及合理性介绍</td><td></td><td></td><td></td><td>5</td></tr>
<tr><td>7</td><td>能正确回答指导老师的问题</td><td></td><td></td><td></td><td>15</td></tr>
<tr><td>8</td><td>能在规定的时间内完成任务</td><td></td><td></td><td></td><td>10</td></tr>
<tr><td>9</td><td>能与他人团结协作</td><td></td><td></td><td></td><td>5</td></tr>
<tr><td>10</td><td>做好 5S 管理工作</td><td></td><td></td><td></td><td>10</td></tr>
<tr><td colspan="2">合计</td><td></td><td></td><td></td><td>100</td></tr>
<tr><td colspan="2">拓展项目</td><td></td><td></td><td></td><td></td></tr>
<tr><td colspan="2">总分</td><td></td><td></td><td></td><td></td></tr>
</table>

评分说明：

① 评分序号 3 为“课前准备”部分评分分值。

② 总分 = “自我评价” ×20% + “小组评价” ×20% + “教师评价” ×20% + 拓展项目。

③ 如有拓展项目，每完成一个拓展项目，总分加 10 分。

## 四、总结反思

1. 总结学到的新知识。

2. 你对自己在本次任务中的表现是否满意？写出课后反思。

3. 列出你掌握的新技能。

## 五、拓展项目

发电机输出电压过低故障排除分析（请自行附纸）。

# 项目5　装载机空调系统检修

## 任务　装载机压缩机的检测及更换

### ☞ 任务描述

通过学习，对装载机压缩机进行分析检修。

### ☞ 知识目标

1）掌握装载机压缩机的作用及工作原理。

2）掌握装载机压缩机的结构。

### ☞ 技能目标

能正确描述装载机压缩机的工作原理。

#### 一、应知应会

1）装载机压缩机的作用与工作原理。

2）装载机压缩机结构。

#### 二、工作过程

**（一）课前准备**

为完成该任务，请检验你是否已掌握以下知识或能力。

1. 请写出装载机压缩机的作用与工作原理。

2. 请写出装载机压缩机的结构。

**（二）计划**

1. 小组分工。

工作页表 5-1　小组分工

<table>
<tr><td rowspan="4">小组信息</td><td>班级</td><td colspan="2"></td><td>日期</td><td colspan="2"></td></tr>
<tr><td>小组名称</td><td colspan="2"></td><td>组长</td><td colspan="2"></td></tr>
<tr><td>岗位分工</td><td></td><td></td><td></td><td></td><td></td></tr>
<tr><td>成员</td><td></td><td></td><td></td><td></td><td></td></tr>
</table>

2. 计划讨论。

小组成员共同讨论工作计划，列出本次任务需要做的工作。

### （三）实施

1. 任务实施。

用户反馈一台装载机的空调不制冷，经检查是压缩机损坏了，需要对压缩机进行拆卸更换。

2. 成果分享。

每个小组将实施结果上传到线上教学平台，由 2 ~ 3 个小组分别展示和讲解 PPT。

## 三、评价

小组成员各自完成“自我评价”，组长完成“小组评价”，教师完成“教师评价”。

工作页表 5-2　任务评价表

| 序号 | 评价内容 | 自我评价 | 小组评价 | 教师评价 | 分值分配 |
|---|---|---|---|---|---|
| 1 | 遵守安全操作规范 | | | | 5 |
| 2 | 态度端正，工作认真 | | | | 5 |
| 3 | 能进行课前学习，完成任务信息相关练习 | | | | 20 |
| 4 | 能熟练、多渠道查找参考资料 | | | | 5 |
| 5 | 能正确完成任务 | | | | 20 |
| 6 | 方案优化及合理性介绍 | | | | 5 |
| 7 | 能正确回答指导老师的问题 | | | | 15 |
| 8 | 能在规定的时间内完成任务 | | | | 10 |
| 9 | 能与他人团结协作 | | | | 5 |
| 10 | 做好 5S 管理工作 | | | | 10 |
| 合计 | | | | | 100 |
| 拓展项目 | | | | | |
| 总分 | | | | | |

评分说明：

① 评分序号 3 为“课前准备”部分评分分值。

② 总分 = “自我评价” ×20% + “小组评价” ×20% + “教师评价” ×20% + 拓展项目。

③ 如有拓展项目，每完成一个拓展项目，总分加 10 分。

## 四、总结反思

1. 总结学到的新知识。

2. 你对自己在本次任务中的表现是否满意？写出课后反思。

3. 列出你掌握的新技能。

## 五、拓展项目

压缩机拆解（请自行附纸）。

# 项目 6　装载机传动系统检修

## 任务　传动系统原理分析

### ☞ 任务描述

通过学习，向客户推荐合适的驱动桥。

### ☞ 知识目标

1）掌握变矩器的结构及工作原理。
2）掌握变速器的结构及工作原理。
3）掌握驱动桥的结构及工作原理。

### ☞ 技能目标

1）能正确描述装载机变矩器的工作原理。
2）能正确描述装载机变速器的工作原理。
3）能正确描述装载机驱动桥的工作原理。

### 一、应知应会

1）装载机变矩器的结构与工作原理。
2）装载机变速器的结构与工作原理。
3）装载机驱动桥的结构与工作原理。

### 二、工作过程

**（一）课前准备**

为完成该任务，请检验你是否已掌握以下知识或能力。

1. 请写出装载机变矩器的结构与工作原理。

2. 请写出装载机变速器的结构与工作原理。

3. 请写出装载机驱动桥的结构与工作原理。

### （二）计划

1. 小组分工。

工作页表 6-1 小组分工

<table>
<tr><td rowspan="4">小组信息</td><td>班级</td><td colspan="2"></td><td>日期</td><td colspan="2"></td></tr>
<tr><td>小组名称</td><td colspan="2"></td><td>组长</td><td colspan="2"></td></tr>
<tr><td>岗位分工</td><td></td><td></td><td></td><td></td><td></td></tr>
<tr><td>成员</td><td></td><td></td><td></td><td></td><td></td></tr>
</table>

2. 计划讨论。

小组成员共同讨论工作计划，列出本次任务需要做的工作。

### （三）实施

1. 任务实施。

客户来购买机器，如何更好地向客户介绍驱动桥，以体现专业水准？了解驱动桥的相关知识，以便给客户推荐合适的驱动桥。

2. 成果分享。

每个小组将实施结果上传到线上教学平台，由 2 ~ 3 个小组分别展示和讲解 PPT。

## 三、评价

小组成员各自完成“自我评价”，组长完成“小组评价”，教师完成“教师评价”。

工作页表 6-2 任务评价表

| 序号 | 评价内容 | 自我评价 | 小组评价 | 教师评价 | 分值分配 |
|---|---|---|---|---|---|
| 1 | 遵守安全操作规范 | | | | 5 |
| 2 | 态度端正，工作认真 | | | | 5 |

（续）

| 序号 | 评价内容 | 自我评价 | 小组评价 | 教师评价 | 分值分配 |
| --- | --- | --- | --- | --- | --- |
| 3 | 能进行课前学习，完成任务信息相关练习 | | | | 20 |
| 4 | 能熟练、多渠道查找参考资料 | | | | 5 |
| 5 | 能正确完成任务 | | | | 20 |
| 6 | 方案优化及合理性介绍 | | | | 5 |
| 7 | 能正确回答指导老师的问题 | | | | 15 |
| 8 | 能在规定的时间内完成任务 | | | | 10 |
| 9 | 能与他人团结协作 | | | | 5 |
| 10 | 做好5S管理工作 | | | | 10 |
| 合计 | | | | | 100 |
| 拓展项目 | | | | | |
| 总分 | | | | | |

评分说明：

① 评分序号3为“课前准备”部分评分分值。

② 总分=“自我评价”×20%+“小组评价”×20%+“教师评价”×20%+拓展项目。

③ 如有拓展项目，每完成一个拓展项目，总分加10分。

## 四、总结反思

1. 总结学到的新知识。

2. 你对自己在本次任务中的表现是否满意？写出课后反思。

3. 列出你掌握的新技能。

## 五、拓展项目

装载机变速器工作原理介绍（请自行附纸）。

# Maintenance and Repair Service of Loader

## Worksheets

**Name:** ______________

**Student ID:** ______________

**Class ID:** ______________

机械工业出版社

# Project 1　Introduction, delivery and inspection of the loader

## Task 1.1　Introduction of the loader

### Task Description

To learn the definition, application scope, classification, functions, and roles of a loader.

### Knowledge Objectives

1) To understand the meaning of the loader.

2) To understand the classification of the loader.

3) To master the performance parameters of the loader and the names of the components.

### Skill Objectives

1) To be able to give examples of scenarios using of the loader.

2) To be able to distinguish the types of the loader.

3) To be able to prepare a PPT introducing CLG856 loader.

**Ⅰ. What you should know**

The structure of the loader, the names of the components and the basic performance parameters.

**Ⅱ. Work procedure**

**(Ⅰ) Pre – class preparation**

To complete the task, please check whether you have mastered the following knowledge or skills.

1. Please write down the definition of the loader.

2. Worksheet Fig. 1-1 shows the general structure of the loader, write down the names of the parts.

Worksheet Fig. 1-1 General structure of the loader

1. ____________________ 2. ____________________
3. ____________________ 4. ____________________
5. ____________________ 6. ____________________
7. ____________________ 8. ____________________
9. ____________________ 10. ____________________
11. ____________________ 12. ____________________
13. ____________________ 14. ____________________
15. ____________________ 16. ____________________
17. ____________________ 18. ____________________
19. ____________________ 20. ____________________
21. ____________________

3. Write the performance parameters of the loader.

### (Ⅱ) Planning

1. Grouping and task assigning.

Worksheet Table 1-1 Grouping and task assigning

| Group information | Class | | | Date | | |
|---|---|---|---|---|---|---|
| | Group name | | | Group leader | | |
| | Tasks | | | | | |
| | Members | | | | | |

2. Plan discussion.

The group members discuss the work plan together and list the work that needs to be done for this task.

### (Ⅲ) Implementation

1. Task implementation.

Prepare a PPT to introduce CLG856H loader.

2. Task sharing.

Each group posts the implementation results to the online teaching platform, and 2 or

3 groups make presentations based on the PPT, respectively.

### Ⅲ. Assessment

Each group member should complete "self - assessment", the group leader should complete "group assessment", and the teacher should complete "teacher assessment".

**Worksheet Table 1-2 Form of task assessment**

| No. | Assessment contents | Self - assessment | Group assessment | Teacher assessment | Value |
|---|---|---|---|---|---|
| 1 | Compliance with safety practices | | | | 5 |
| 2 | Work with a good attitude and conscientiousness | | | | 5 |
| 3 | To be able to study in advance of class and complete the exercises related to the task information | | | | 20 |
| 4 | To be able to find reference materials in a variety of manners proficiently | | | | 5 |
| 5 | To be able to complete the tasks correctly | | | | 20 |
| 6 | Optimized program with reasonable presentation | | | | 5 |
| 7 | To be able to answer the instructor's questions correctly | | | | 15 |
| 8 | To be able to complete tasks within the time limit | | | | 10 |
| 9 | To be able to cooperate with others | | | | 5 |
| 10 | To carry out 5S management strictly | | | | 10 |
| Total | | | | | 100 |
| Extended Project | | | | | |
| Total Score | | | | | |

Notes on scoring:

① Item 3 is the score of "Preparation for class".

② Total score = "Self - assessment" ×20% + "Group assessment" ×20% + "Teacher assessment" ×20% + Extended Project.

③ If there is an extended project, the total score will be increased by 10 points for each completed Extended Project.

### Ⅳ. Summary and reflection

1. The new knowledge you have learned.

2. Are you satisfied with your performance in this task? Write a post – class reflection.

## V. Extended project

Write down a list of the components and names of each part of the loader. (Please attach your own paper.)

# Task 1.2 Delivery and inspection of the loader

## Task Description

How to complete the delivery and inspection of a new loader to be delivered to the customer?

## Knowledge Objectives

1) To master the inspection contents and standards of the loader delivery.

2) To master the precautions of operation and storage of the loader.

## Skill Objectives

1) To be able to list the contents of the loader delivery inspection.

2) To be able to prepare a delivery inspection report of the CLG856 loader.

**Ⅰ. What you should know**

Contents and standards of the loader for delivery and inspection

**Ⅱ. Work procedure**

**(Ⅰ) Pre-class Preparation**

To complete the task, please check whether you have mastered the following knowledge or ability.

1. Please list the key inspection parts of the loader

Worksheet Table 1-3 The key inspection parts

| No. | Acceptance items | Result | | No. | Acceptance items | Result | |
|---|---|---|---|---|---|---|---|
| | | Normal | Abnormal | | | Normal | Abnormal |
| 1 | | | | 7 | | | |
| 2 | | | | 8 | | | |
| 3 | | | | 9 | | | |
| 4 | | | | 10 | | | |
| 5 | | | | 11 | | | |
| 6 | | | | 12 | | | |

2. Please explain the contents of the whole loader inspection and how to test them respectively?

3. Work out the precautions for operation and storage of the whole machine.

**(Ⅱ) Planning**

1. Grouping and task assigning.

Worksheet Table 1-4 Grouping and task assigning

| Group information | Class | | | Date | | |
|---|---|---|---|---|---|---|
| | Group name | | | Group leader | | |
| | Tasks | | | | | |
| | Members | | | | | |

2. Plan discussion.

The group members discuss the work plan together and list the work that needs to be done for this task.

### (Ⅲ) Implementation

1. Task implementation.

Be able to compile a loader delivery and inspection report.

2. Task sharing.

Each group posts their implementation results to the online teaching platform, and 2 or 3 groups make presentations and explain the PPT, respectively.

## Ⅲ. Assessment

Each group member should complete "self – assessment", the group leader should complete "group assessment", and the teacher should complete "teacher assessment".

Worksheet Table 1-5 Form of task assessment

| No. | Assessment contents | Self – assessment | Group assessment | Teacher assessment | Value |
|---|---|---|---|---|---|
| 1 | Compliance with the safety practices | | | | 5 |
| 2 | Work with a good attitude and conscientiousness | | | | 5 |
| 3 | To be able to study in advance of class and complete exercises related to task information | | | | 20 |
| 4 | To be able to find references in a variety of efficient ways | | | | 5 |
| 5 | To be able to complete tasks correctly | | | | 20 |
| 6 | To be able to optimize programs with rationalized interpretation | | | | 5 |
| 7 | To be able to answer the instructor's questions correctly | | | | 15 |
| 8 | To be able to complete tasks within the time limit | | | | 10 |
| 9 | To be able to cooperate with others | | | | 5 |
| 10 | To carry out 5S management strictly | | | | 10 |
| Total | | | | | 100 |
| Extended Project | | | | | |
| Total Score | | | | | |

Notes on scoring:

① Item 3 is the score of "Preparation for class".

② Total score = "Self – assessment" ×20% + "Group assessment" ×20% + "Teacher assessment" ×20% + Extended Project.

③ If there is an extended project, the total score will be increased by 10 points for each Extended Project completed.

## Ⅳ. Summary and reflection

1. The new knowledge you have learned.

2. Are you satisfied with your performance in this task? Write a post - class reflection.

## Ⅴ. Extended Project

Write out the key delivery inspection parts of the loader. (Please attach your own paper.)

# Project 2  Maintenance of the loader

## Task  Preparation of regular maintenance and service plans

### Task Description

By learning, make a 1000h regular maintenance plan for the CLG856H loader.

### Knowledge Objectives

1) To understand the contents and requirements of the periodic maintenance of the loader.

2) To be able to know the preparation of a regular maintenance plan for a loader.

### Skill Objectives

To be able to prepare regular maintenance content for the CLG856H loader.

**Ⅰ. What you should know**

The contents and requirements of the regular maintenance.

**Ⅱ. Work Procedure**

**(Ⅰ) Pre - class preparation**

To complete the task, please check whether you have mastered the following knowledge or abilities.

1. Please write down the items of the loader to be checked before running - in.

2. Please write down the items that need to be inspected for the running - in loader.

3. Please write down the items that need to be maintained regularly for the loaders.

**(Ⅱ) Planning**

1. Grouping and task assigning.

**Worksheet Table 2-1　Grouping and task assigning**

| Group information | Class | | | Date | | |
|---|---|---|---|---|---|---|
| | Group name | | | Group leader | | |
| | Tasks | | | | | |
| | Members | | | | | |

2. Plan discussion.

The group members discuss the work plan together and list the work that needs to be done for this task.

**(Ⅲ) Implementation**

1. Task implementation.

To complete the preparation of the regular maintenance schedule for the CLG856H loader.

2. Task sharing.

Each group posts their implementation results to the online teaching platform, and 2 or 3 groups make presentations based on the PPT, respectively.

**Ⅲ. Assessment**

Each group member should complete "self－assessment", the group leader should complete "group assessment", and the teacher should complete "teacher assessment".

**Worksheet Table 2-2　Form of task assessment**

| No. | Assessment contents | Self－assessment | Group assessment | Teacher assessment | Value |
|---|---|---|---|---|---|
| 1 | Compliance with safety practices | | | | 5 |
| 2 | Work with a good attitude and conscientiousness | | | | 5 |
| 3 | To be able to study in advance of class and complete exercises related to task information | | | | 20 |
| 4 | To be able to find reference materials in a proficient and multi－channel manner | | | | 5 |
| 5 | To be able to complete tasks correctly | | | | 20 |
| 6 | To be able to optimize programs with rationalized interpretation | | | | 5 |
| 7 | To be able to answer the instructor's questions correctly | | | | 15 |
| 8 | To be able to complete tasks within the time limit | | | | 10 |
| 9 | To be able to cooperate with others | | | | 5 |
| 10 | To carry out 5S management strictly | | | | 10 |

(续)

| No. | Assessment contents | Self - assessment | Group assessment | Teacher assessment | Value |
|---|---|---|---|---|---|
| Total | | | | | 100 |
| Extended Project | | | | | |
| Total Score | | | | | |

Notes on scoring:

① Item 3 is the score of "Preparation for class".

② Total score = "Self - assessment" ×20% + "Group assessment" ×20% + "Teacher assessment" ×20% + Extended project.

③ If there is an extended project, the total score will be increased by 10 points for each extended project completed.

## Ⅳ. Summary and reflection

1. The new knowledge you have learned.

2. Are you satisfied with your performance in this task? Write a post - class reflection.

## Ⅴ. Extended Project

Prepare a 1000h regular maintenance schedule for the loader. (Please attach your own paper.)

# Project 3　Inspecting the hydraulic system of the loader

## Task 3.1　Analysing the working principle of the hydraulic system

### Task Description

By learning, complete the overhaul of the hydraulic system of a loader.

### Knowledge Objectives

1) Master the control circuit principles of the pilot hydraulic system.

2) Master the principles of the pilot circuit, analyse and eliminate common faults.

3) Master the control circuit principles of the bucket hydraulic system.

4) Master the principles of the bucket circuit, analyze and eliminate common faults.

5) Master the control circuit principles of the boom hydraulic system.

6) Master the principles of the boom circuit, analyse and eliminate common faults.

7) Master the control circuit principles of the steering hydraulic system.

8) Master the common steering hydraulic system faults and make analysis and elimination.

9) Master the control circuit principles of the braking system.

10) Master the common braking system faults and make analysis and elimination.

### Skill Objectives

1) To be able to analyse the working hydraulic system control circuit.

2) To be able to analyse and troubleshoot the common faults through the principle of control circuit.

**Ⅰ. What you should know**

1) The working principles of the loader.

2) Composition of the loader hydraulic system.

3) The working principle of the pilot hydraulic system.

4) The working principle of the bucket hydraulic circuit.

5) The working principle of the hydraulic circuit of the boom.

6) The working principle of the steering system control circuit.

7) The working principle of the control circuit of the braking system.

8) The pressure test of the working hydraulic system.

## Ⅱ. Work procedure

### (Ⅰ) Pre – class Preparation

To complete the task, please check whether you have mastered the following knowledge or abilities.

1. Please write down the working principle of the loader hydraulic system.

2. Please write down the compositions of the loader hydraulic system.

3. Please write down the pilot hydraulic system, bucket hydraulic system, and the boom hydraulic system circuit of the loader.

4. Please write down the pressure test process of the hydraulic system of the loader.

5. Please write down the pressure test process of the steering hydraulic system of the loader.

6. Please write down the pressure test process of the braking system of the loader.

### (Ⅱ) Planning

1. Grouping and task assigning.

**Worksheet Table 3-1 Grouping and task assigning**

| Group information | Class | | | Date | | |
|---|---|---|---|---|---|---|
| | Group name | | | Group leader | | |
| | Tasks | | | | | |
| | Members | | | | | |

2. Plan discussion.

The group members discuss the work plan together and list the work that needs to be done for this task.

**(Ⅲ) Implementation**

1. Task implementation.

If the boom of a loader has a weak lift, and the tested system pressure is less than the designed value, then the system pressure needs to be tested and adjusted.

2. Task sharing.

Each group is required to upload the implementation results to the online teaching platform, and 2 or 3 groups are required to make presentations based on the PPT, respectively.

## Ⅲ. Assessment

Each group member should complete "self – assessment", the group leader should complete "group assessment", and the teacher should complete "teacher assessment".

**Worksheet Form of task assessment**

| No. | Assessment contents | Self – assessment | Group assessment | Teacher assessment | Value |
|---|---|---|---|---|---|
| 1 | Compliance with safety practices | | | | 5 |
| 2 | Work with a good attitude and conscientiousness | | | | 5 |
| 3 | To be able to study in advance of class and complete exercises related to task information | | | | 20 |
| 4 | To be able to find references in a variety of efficient ways | | | | 5 |
| 5 | To be able to complete tasks correctly | | | | 20 |
| 6 | To be able to optimize programs with rationalized interpretation | | | | 5 |
| 7 | To be able to answer the instructor's questions correctly | | | | 15 |
| 8 | To be able to complete tasks within the time limit | | | | 10 |
| 9 | To be able to cooperate with others | | | | 5 |

(续)

| No. | Assessment contents | Self – assessment | Group assessment | Teacher assessment | Value |
|---|---|---|---|---|---|
| 10 | To carry out 5S management strictly | | | | 10 |
| Total | | | | | 100 |
| Extended project | | | | | |
| Total score | | | | | |

Notes on scoring:

① Item 3 is the score of "Preparation for class".

② Total score = "Self – assessment" ×20% + "Group assessment" ×20% + "Teacher assessment" ×20% + Extended project.

③ If there is an extended project, the total score will be increased by 10 points for each extended project completed.

## Ⅳ. Summary and reflection

1. The new knowledge you have learned.

2. Are you satisfied with your performance in this task? Write a post – class reflection.

3. List the new skills you have mastered.

## Ⅴ. Extended project

Testing the pressure of the steering system of the loader. (Please attach your own paper.)

# Task 3.2 Testing the performance of the hydraulic system

## Task Description

To complete the performance testing and measurement of the hydraulic system of loader.

## Knowledge Objectives

1) Master the test for loader hydraulic cylinder settlement.
2) Master the test for loader hydraulic cylinder cycle time.
3) Master the test for loader hydraulic oil temperature.

## Skill Objectives

1) To be able to measure the hydraulic cylinder settlement of the loader according to the specification requirements.

2) To be able to measure the hydraulic cylinder cycle time according to the specification requirements.

3) To be able to use the temperature gun to measure the hydraulic oil temperature.

**Ⅰ. What you should know**

(1) Testing of loader hydraulic cylinder settlement.

(2) Testing of loader hydraulic cylinder cycle time.

(3) Testing of loader hydraulic oil temperature.

**Ⅱ. Work procedure**

**(Ⅰ) Pre – class preparation**

To complete the task, please check whether you have mastered the following knowledge or abilities.

1. Please write a plan for testing the hydraulic cylinder of the loader.

2. Please write the steps of testing the hydraulic cylinder cycle time of the loader.

**(Ⅱ) Planning**

1. Grouping and task assigning.

Worksheet Table 3-3 Grouping and task assigning

<table>
<tr><td rowspan="4">Group information</td><td>Class</td><td colspan="2"></td><td>Date</td><td colspan="2"></td></tr>
<tr><td>Group name</td><td colspan="2"></td><td>Group leader</td><td colspan="2"></td></tr>
<tr><td>Tasks</td><td></td><td></td><td></td><td></td><td></td></tr>
<tr><td>Members</td><td></td><td></td><td></td><td></td><td></td></tr>
</table>

2. Plan discussion.

The group members discuss the work plan together and list the work that needs to be done for this task.

### (Ⅲ) Implementation

1. Task iImplementation.

Suppose when a loader lifts to the highest level, the bucket drops. In this case, it is necessary to test the settlement of the hydraulic cylinder to find the cause.

2. Task sharing.

Each group is required to upload the implementation results to the online teaching platform, and 2 or 3 groups are required to make presentations based on the PPT, respectively.

## Ⅲ. Evaluation

Each group member should complete "self – assessment", the group leader should complete "group assessment", and the teacher should complete "teacher assessment".

Worksheet Table 3-4 Form of task assessment

| No. | Assessment contents | Self – assessment | Group assessment | Teacher assessment | Value |
|---|---|---|---|---|---|
| 1 | Compliance with safety practices | | | | 5 |
| 2 | Work with a good attitude and conscientiousness | | | | 5 |
| 3 | To be able to study in advance of class and complete exercises related to the task | | | | 20 |
| 4 | To be able to find references in a variety of efficient ways | | | | 5 |
| 5 | To be able to complete tasks correctly | | | | 20 |
| 6 | To be able to optimize programs with rationalized interpretation | | | | 5 |
| 7 | To be able to answer the instructor's questions correctly | | | | 15 |
| 8 | To be able to complete tasks within the time limit | | | | 10 |
| 9 | To be able to cooperate with others | | | | 5 |
| 10 | To carry out 5S management strictly | | | | 10 |
| Total | | | | | 100 |
| Extended Project | | | | | |
| Total Score | | | | | |

Notes on scoring:

① Item 3 is the score of "Preparation for class".

② Total score = "Self – assessment" × 20% + "Group assessment" × 20% + "Teacher assessment" × 20% + Extended project.

③ If there is an extended project, the total score will be increased by 10 points for each extended project completed.

## Ⅳ. Summary and reflection

1. The new knowledge you have learned.

2. Are you satisfied with your performance in this task? Write a post - class reflection.

3. list the new skills you have mastered.

## Ⅴ. Extended project

Testing hydraulic oil temperature of loaders. (Please attach your own paper.)

# Project 4　Maintenance of an electrical system of the loader

## Task　Analysis of the working principle of main circuits

### Task Description

To complete the overhaul of the hydraulic system of a loader and to analyse and discriminate the working principle.

### Knowledge Objectives

1) To master the structure and functions of the main circuits.

2) Master the analysis of the main parts of the main circuit.

3) To master the working principle of the main circuits and the analysis of faults.

### Skill Objectives

1) To be able to correctly describe the working principle of the main circuit of the loader.

2) To be able to overhaul the power circuit according to the specifications.

**Ⅰ. What you should know**

1) The structure and functions of the main circuit.

2) The main parts of the main circuit of the loader.

3) The working principle of the main circuits and the analysis of faults.

**Ⅱ. Work Procedure**

**(Ⅰ) Pre – class preparation**

To complete the task, please check whether you have mastered the following knowledge or abilities.

1. Please write down the main structure of the electrical system of a loader.

2. Please write down the working principle of the main circuits.

3. Please analyse the faults by using the method of listing.

### (Ⅱ) Planning

1. Grouping and task assigning.

Worksheet Table 4-1 Grouping and task assigning

<table>
<tr><td rowspan="4">Group information</td><td>Class</td><td colspan="2"></td><td>Date</td><td colspan="2"></td></tr>
<tr><td>Group name</td><td colspan="2"></td><td>Group leader</td><td colspan="2"></td></tr>
<tr><td>Tasks</td><td></td><td></td><td></td><td></td><td></td></tr>
<tr><td>Members</td><td></td><td></td><td></td><td></td><td></td></tr>
</table>

2. Plan discussion.

The group members discuss the work plan together and list the work that needs to be done for this task.

### (Ⅲ) Implementation

1. Task implementation.

The generator of the loader is blocked up with collision and scraping noise. It is necessary to inspect and analyse the generator and find solutions.

2. Task sharing.

Each group is required to upload the implementation results to the on – line teaching platform, and 2 or 3 groups are required to make presentations based on the PPT, respectively.

## Ⅲ. Assessment

Each group member should complete "self – assessment", the group leader should complete "group assessment", and the teacher should complete "teacher assessment".

Worksheet Table 4-2 Task assessment

| No. | Assessment contents | Self – assessment | Group assessment | Teacher assessment | Value |
|---|---|---|---|---|---|
| 1 | Compliance with safety practices | | | | 5 |
| 2 | Work with good attitude and conscientiousness | | | | 5 |
| 3 | To be able to study in advance of class and complete exercises related to the task | | | | 20 |

(续)

| No. | Assessment contents | Self - assessment | Group assessment | Teacher assessment | Value |
|---|---|---|---|---|---|
| 4 | To be able to find references in a variety of efficient ways | | | | 5 |
| 5 | To be able to complete tasks correctly | | | | 20 |
| 6 | To be able to optimize programs with rationalized interpretation | | | | 5 |
| 7 | To be able to answer the instructor's questions correctly | | | | 15 |
| 8 | To be able to complete tasks within the time limit | | | | 10 |
| 9 | To be able to cooperate with others | | | | 5 |
| 10 | To carry out 5S management strictly | | | | 10 |
| Total | | | | | 100 |
| Extended project | | | | | |
| Total score | | | | | |

Notes on scoring

① Item 3 is the score of "Preparation for class".

② Total score = "Self - assessment" × 20% + "Group assessment" × 20% + "Teacher assessment" × 20% + Extended project.

③ If there is an extended project, the total score will be increased by 10 points for each extended project completed.

## Ⅳ. Summary and Reflection

1. The new knowledge you have learned.

2. Are you satisfied with your performance in this task? Write a post - class reflection.

3. list the new skills you have mastered.

## V. Extended Project

Troubleshooting and analysing on the low generator output voltage. ( Please attach your own paper. )

# Project 5　Maintenance of the air conditioning system of the loader

## Task　Compressor testing and replacement

### Task Description

Analysis and overhaul of the air conditioning compressor of the loader.

### Knowledge Objectives

1) To master the roles and working principle of the compressor.

2) To master the structure of the compressor.

### Skill Objectives

To be able to correctly describe the working principle of the air conditioning compressor of the loader.

**Ⅰ. What you should know**

1) The role and working principle of the loader compressor.

2) The structure of the compressor of the loader.

**Ⅱ. Work procedure**

**(Ⅰ) Pre – class preparation**

To complete the task, please check whether you have mastered the following knowledge or abilities.

1. Please write down the roles and working principles of the compressor of the loader.

2. Please write down the structure of the compressor of the loader.

**(Ⅱ) Planning**

1. Grouping and task assigning.

Worksheet Table 5-1 Grouping and task assigning

<table>
<tr><td rowspan="4">Group information</td><td>Class</td><td colspan="2"></td><td>Date</td><td colspan="2"></td></tr>
<tr><td>Group name</td><td colspan="2"></td><td>Group leader</td><td colspan="2"></td></tr>
<tr><td>Tasks</td><td></td><td></td><td></td><td></td><td></td></tr>
<tr><td>Members</td><td></td><td></td><td></td><td></td><td></td></tr>
</table>

2. Plan discussion.

The group members discuss the work plan together and list the work that needs to be done for this task.

### (Ⅲ) Implementation

1. Task implementation.

A loader user feedback shows that the air conditioning is not cooling. After inspection, it is found that the compressor is damaged, so the compressor needs to be disassembled and replaced.

2. Task sharing.

Each group is required to upload the implementation results to the online teaching platform, and 2 or 3 groups are required to make presentations based on the PPT, respectively.

## Ⅲ. Assessment

Each group member should complete "self – assessment", the group leader should complete "group assessment", and the teacher should complete "teacher assessment".

Worksheet Table 5-2 Task assessment

| No. | Assessment contents | Self – assessment | Group assessment | Teacher assessment | Value |
|---|---|---|---|---|---|
| 1 | Compliance with safety practices | | | | 5 |
| 2 | Work with good attitude and conscientiousness | | | | 5 |
| 3 | To be able to study in advance of class and complete exercises related to the task | | | | 20 |
| 4 | To be able to find references in a variety of efficient ways | | | | 5 |
| 5 | To be able to complete tasks correctly. | | | | 20 |
| 6 | To be able to optimize programs with rationalized interpretation | | | | 5 |
| 7 | To be able to answer the instructor's questions correctly | | | | 15 |
| 8 | To be able to complete tasks within the time limit | | | | 10 |
| 9 | To be able to cooperate with others | | | | 5 |

(续)

| No. | Assessment contents | Self - assessment | Group assessment | Teacher assessment | Value |
|---|---|---|---|---|---|
| 10 | To carry out 5S management strictly | | | | 10 |
| | Total | | | | 100 |
| | Extended project | | | | |
| | Total score | | | | |

Notes on scoring

① Item 3 is the score of "Preparation for class".

② Total score = "Self - assessment" ×20% + "Group assessment" × 20% + "Teacher assessment" ×20% + Extended project.

③ If there is an extended project, the total score will be increased by 10 points for each extended project completed.

## Ⅳ. Summary and reflection

1. The new knowledge you have learned.

2. Are you satisfied with your performance in this task? Write a post - class reflection.

3. List the new skills you have mastered.

## Ⅴ. Extended project

Disassemble a compressor. (Please attach your own paper.)

# Project 6 Inspection and repair the drive system of the loader

## Task Analysing the principle of transmission system

### Task Description

By learning, recommend suitable drive axles to customers.

### Knowledge Objectives

1) To master the structure and operation principle of a torque converter.

2) To master the structure and operation principle of the gearbox.

3) To master the structure and operation principle of the driving axle.

### Skill Objectives

1) Be able to correctly describe the operation principle of the torque converter of the loader.

2) Be able to correctly describe the operation principle of the loader gearbox.

3) Be able to correctly describe the operation principle of the drive axle of the loader.

**I. What you should know**

1) The structure and operation principle of the torque converter of the loader.

2) The structure and operation principle of the loader gearbox.

3) The structure and operation principle of the drive axle of the loader.

**II. Work procedure**

**(I) Pre-class preparation**

To complete the task, please check whether you have mastered the following knowledge or abilities.

1. Please write down the structure and operation principle of the torque converter of the loader.

2. Please write down the structure and operation principle of transmissions of the loader.

3. Please write down the structure and operation principle of a drive axle of the loader.

### (Ⅱ) Planning

1. Grouping and task assigning

Worksheet Table 6-1 Grouping and task assigning

<table>
<tr><td rowspan="4">Group information</td><td>Class</td><td colspan="2"></td><td>Date</td><td colspan="2"></td></tr>
<tr><td>Group name</td><td colspan="2"></td><td>Group leader</td><td colspan="2"></td></tr>
<tr><td>Tasks</td><td></td><td></td><td></td><td></td><td></td></tr>
<tr><td>Members</td><td></td><td></td><td></td><td></td><td></td></tr>
</table>

2. Plan discussion

The group members discuss the work plan together and list the work that needs to be done for this task.

### (Ⅲ) Implementation

1. Task implementation

When a customer comes to buy a mechine, how to better introduce the drive axle to him/her in a professional way? You need to understand relevant knowledge of drive axles in order to recommend the most suitable one to the customer.

2. Task sharing

Each group is required to upload the implementation results to the on – line teaching platform, and 2 or 3 groups are required to make presentations based on the PPT, respectively.

## Ⅲ. Assessment

Each group member should complete "self – assessment", the group leader should complete "group assessment", and the teacher should complete "teacher assessment".

**Worksheet Table 6-2 Task Assessment**

| No. | Assessment contents | Self – assessment | Group assessment | Teacher assessment | Value |
|---|---|---|---|---|---|
| 1 | Compliance with safety practices | | | | 5 |
| 2 | Work with good attitude and conscientiousness | | | | 5 |
| 3 | To be able to study in advance of class and complete exercises related to the task | | | | 20 |
| 4 | To be able to find references in a variety of efficient ways | | | | 5 |
| 5 | To be able to complete tasks correctly | | | | 20 |
| 6 | To be able to optimize programs with rationalized interpretation | | | | 5 |
| 7 | To be able to answer the instructor's questions correctly | | | | 15 |
| 8 | To be able to complete tasks within the time limit | | | | 10 |
| 9 | To be able to cooperate with others | | | | 5 |
| 10 | To carry out 5S management strictly | | | | 10 |
| Total | | | | | 100 |
| Extended Project | | | | | |
| Total Score | | | | | |

Notes on scoring

① Item 3 is the score of "Preparation for class".

② Total score = "Self – assessment" ×20% + "Group assessment" ×20% + "Teacher assessment" ×20% + Extended project.

③ If there is an extended project, the total score will be increased by 10 points for each extended project completed.

## Ⅳ. Summary and reflection

1. The new knowledge you have learned.

2. Are you satisfied with your performance in this task? Write a post – class reflection.

3. list the new skills you have mastered.

## V. Extended Project

Introducing the structure and operation principle of the gearbox of a loader. (Please attach your own paper.)